SEEDS

by Melanie Mitchell

first step nonfiction

Lerner Publications Company · Minneapolis

Look at the seeds.

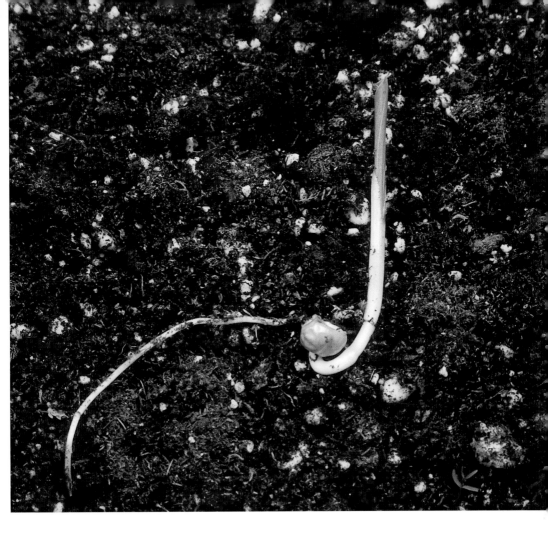

Some plants grow from
seeds.

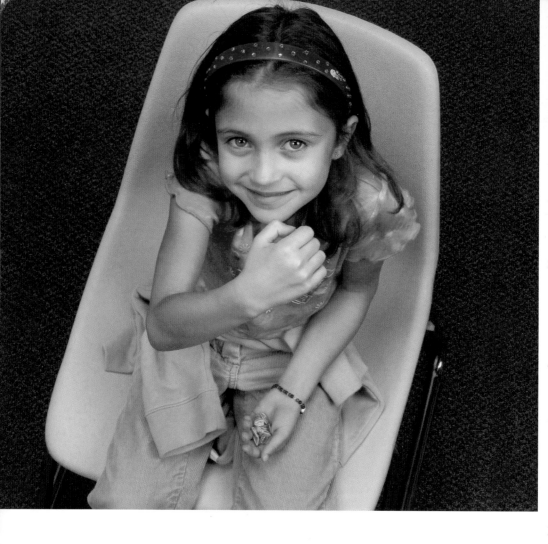

Sometimes people eat seeds.

Seeds are many colors.

Seeds can be small.

Seeds can be big.

We like to plant seeds.